TIME

Pamela J.P. Schroeder
Jean M. Donisch

ROURKE PUBLICATIONS, INC.
VERO BEACH, FL 32964

A book by Market Square Communications Incorporated.
A special thanks to our creative team, Sandy Robinson, Sandra Shekels and
Ann Garber of Market Square Communications Incorporated, for their creative
text and design contributions.

Consultants:
        Jeanette Handrich — M.A. in Elementary Education/Language Arts,
           third and fourth grade teacher/gifted and talented program,
           over 20 years teaching experience
        Karen M. Olsen — M.S. in Education, kindergarten teacher, over
           20 years teaching experience
        Geri Pape — M.S. in Elementary Education, kindergarten teacher,
           over 30 years teaching experience

Library of Congress Cataloging-in-Publication Data
Schroeder, Pamela J. P., 1969-
  Time / Pamela J. P. Schroeder, Jean M. Donisch.
    p. cm. — (What's the big idea?)
  Summary: Illustrations and rhyming text introduce various
aspects of the concept of time, such as minutes and hours, day
and night, months and years, young and old. Includes related
questions.
  ISBN 0-86625-578-8 (alk.paper)
  1. Time—Juvenile literature. 2. Night—Juvenile literature.
3. Day—Juvenile literature. [1. Time.] I. Donisch, Jean M.,
1960- . II. Title III. Series.
QB209.5.S37 1996
529—dc20   96-714
                                 CIP
                                 AC

Printed in the U.S.A.

TABLE OF CONTENTS

For more fun
with TIME ideas,
look for this shape
on the page.

WHAT'S THE BIG IDEA? ABOUT TIME

Tick tock, tick tock,
Time keeps moving around the clock.

It's time to get up.
It's time to eat.
It's time to go home.
It's time to sleep.

It's nothing that you can hear or see,
Or anything you can bump with your knee.

Time is a great invisible friend
That tells when things start and when they end.

Time tells us when to work or play,
And when to wait for another day.

Time brings us birthdays and holidays, too,
Weekends and memories, the first day of school.

Turn the page now, and we'll begin
To learn about time, our invisible friend.

What Time Is It?

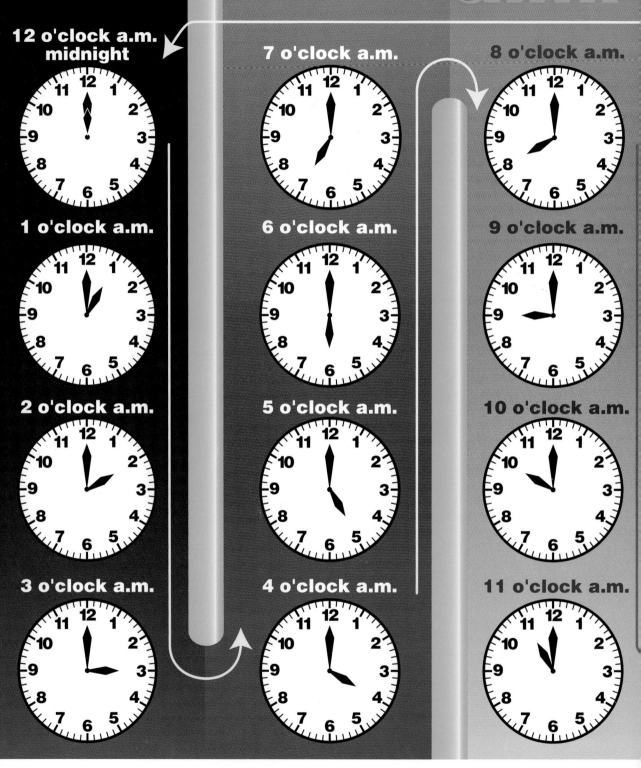

12 o'clock a.m.
midnight

1 o'clock a.m.

2 o'clock a.m.

3 o'clock a.m.

7 o'clock a.m.

6 o'clock a.m.

5 o'clock a.m.

4 o'clock a.m.

8 o'clock a.m.

9 o'clock a.m.

10 o'clock a.m.

11 o'clock a.m.

Twenty-four hours for each day and night.
Watch the little hand go around the clock twice.

p.m.

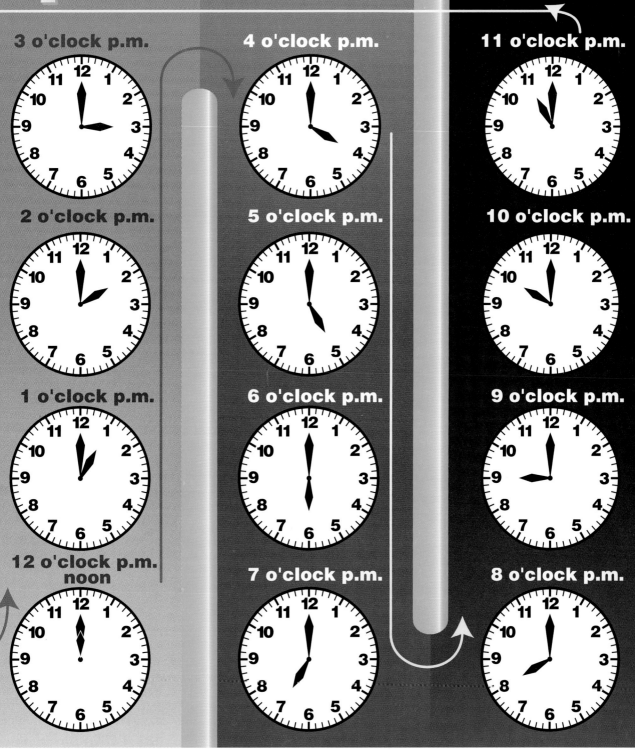

3 o'clock p.m.

4 o'clock p.m.

11 o'clock p.m.

2 o'clock p.m.

5 o'clock p.m.

10 o'clock p.m.

1 o'clock p.m.

6 o'clock p.m.

9 o'clock p.m.

12 o'clock p.m. noon

7 o'clock p.m.

8 o'clock p.m.

Twelve a.m. hours last from midnight to noon.
Twelve p.m. hours last from noon to midnight.

6:30 a.m.
Ring!
It's six-thirty.
Wake up —
get out of bed.
Time to eat,
brush your teeth —
the whole day waits ahead.

10:00 a.m.
Climb, run,
jump and play.
Ten o'clock
recess is the best
time of day.

12:00 p.m.
Twelve noon —
it's time for lunch.
Eat a bunch —
chew and crunch!

8

1:30 p.m.
Is it nap time?
Let's check and see.
One-thirty — music class.
Sing do-re-mi.

What will you be doing tomorrow at 11:00 a.m.?

8:30 p.m.
Reading is fun
and good for your head.
Eight-thirty's a good time
for a story before bed.

6:00 p.m.
Clink, clank
as you put down a plate.
Help set the table
or dinner will be late.

BEDTIME STORIES

9

Seconds

It takes one second ...

... for the long, thin second hand to go from the green line to the orange line.

... to crack an egg.

... to kick a soccer ball.

How many seconds can you stand on one foot?

Minutes

It takes one minute ...

... for the long, thin second hand to go all the way around.

... to beat eggs for a cake.

... to tie your soccer shoes.

Hours

It takes one hour ...

... for the long minute hand to go all the way around the clock.

... to build a sand castle.

1 hour

... to watch a 60-minute show.

... to bake a cake.

... to play a soccer game.

When can an hour seem like a minute?

= 60 minutes

Day

Daytime
is a time for fun.
Wake up
and get out
under the sun.
Time for school
and work and play —
twelve hours is all
you need for a day.

At night the moon
and stars come out.
Big owls with shining eyes
fly about.
It's time to curl up
on your pillow —
sleep tight.
Twelve hours
of quiet —
that makes a night.

If you were a bat, what would be your favorite time?

Night

Seven days in a week – come take a peek.

SUNDAY JUNE 1

TUESDAY JUNE 3

Next it's Tuesday — come and play.

MONDAY JUNE 2

WEDNESDAY JUNE 4

It's Sunday first, let's fly a kite.

On Monday, come let's ride a bike.

Now it's Wednesday — a rainy day.

Week

Saturday finishes our week. Let's play a little hide and seek!

On Thursday, we can sail a ship.

SATURDAY JUNE 7

THURSDAY JUNE 5

RKE1970

FRIDAY JUN 6

What makes Saturday and Sunday different from the rest of the week?

On Friday, let's take a long trip.

Day Thursday Date June 5
Yesterday my tooth was loose. Today it fell out! Tomorrow the tooth fairy will come.

Day Friday Date June 6

Date

Date

17

Month

There are
30 days
in a month,
sometimes 31.
Except
February — with
28 or 29 — is a
special one.

January, February,
March, April, May,
June, July, August,
September, October,
November, December.

Put them all together
and you have a year.
Wait 12 months
and your next birthday
is here!

What is your
favorite
month of
the year?
Why?

EMBER

WEDNESDAY	THURSDAY	FRIDAY	SATURDAY
	Gym Day	Steven Point's Birthday 6	Movie Day 1 p.m. 7
4	5		
Field Trip Children's Museum 11	Gym Day 12	Show And Tell 13	Soccer 10 a.m. 14
	Gym Day 19	Show And Tell Bring nest! 20	Movie Day 1 p.m. 21
17	18		
24	Annie Buddy's Birthday 25		Soccer 10 a.m. 28

Cedaredge Public Library
P.O. Box 548
Cedaredge, CO 81413

19

Year

A year is 12 months,
or 365 days.
Next year's always coming,
but it never stays.

Rain splashes down
as snow melts away.
Spring is a time
for beginnings, they say.

Summer heats up
for picnics and beaches.
No school — outdoor lessons
is what summer teaches!

How many hours are there in a year?

Then pumpkins grow round
and the leaves start to drop.
It's fall. School starts
and Jack Frost starts to hop.

In winter, up north,
sparkling snow starts to fly.
Long nights are cold,
but holidays keep hopes high.

A year brings four seasons —
winter, spring, summer and fall.
Just look out your window
and you'll see it all.

Before

A cookie begins
far away from your kitchen.
Before cookies, you need a cow
and a chicken.

Before you can bake cookies, you
need all these things
that make them taste great and
make your mouth sing.

Before you eat cookies
your mouth says, "Yum!"

After

Chickens lay eggs and milk comes from a cow. After that you can make a cookie — here's how.

After you bake, you'll have cookies galore. But they're hot — wait just a little bit more.

What happens to a cookie after you eat it?

After you eat them, there's nothing but crumbs.

Young

Young is how things are when they've just begun.
A puppy, a kitten, a baby are some.

Old is how things are after time passes —
A fully-grown dog, grandma and grandpa with glasses.

Special

Some special times
aren't times that clocks tell.

They have special names
that we know very well.

Lunchtime fills up our tummies.

Birthdays bring gifts and fun.

Bedtime is for stories.

Holidays cheer
everyone.

What do you like to do with your free time?

Times

But my favorite
is free time —
there's no have-to-dos.

That means I can fill it
with what I want to.

WHAT'S THE BIG IDEA? ABOUT TIME

Tick tock, tick tock,
Time keeps moving
Around the clock.

Time has brought us
To the end of this book.
Now let's stop and take another look
At the way time, our invisible friend,
Touches our lives from beginning to end.

Seconds make minutes
And minutes make hours.
Hours make days and nights
With their powers.

Seven days in a week —
A month needs about 30.
Twelve months make a year, too.
What else can time do?
Time lets us know
What is now, then and when.
Time keeps going for us,
Our invisible friend.

GLOSSARY

a.m. – the 12 hours from midnight to noon

day – made up of 24 hours – 12 daytime hours and 12 nighttime hours

hour – made up of 60 minutes; 24 hours make a day

hour hand – the short hand on clocks that goes around once every 12 hours; it points to the hour of time — 12:00

leap year – every four years, a year with 366 days

midnight – 12:00 a.m.

minute – made up of 60 seconds; 60 minutes make an hour

minute hand – the long hand on clocks that goes around once every hour; it points to the minutes of time — 12:15

month – made up of about 30 days (January, February and so on); 12 months make a year

noon – 12:00 p.m.

p.m. – the 12 hours from noon to midnight

seasons – spring, summer, winter and fall

second – a very short time; 60 seconds make a
    minute

second hand – the long, thin hand on some
    clocks that goes around once every minute

week – made up of seven days (Monday,
    Tuesday and so on); 52 weeks make a year

year – made up of 12 months, 52 weeks or
    365 days

ABOUT TIME

What is a 24-hour clock, and who uses one?

How many minutes does it take you to get home from school? How many seconds would that be?

When can a short time seem like a long time?

What's good about being young? What's good about being old?

What things do you do before school? How about after?

Make up a song or poem to teach someone the days of the week.

How could you tell time if you didn't have a clock?

Try keeping a diary of the things that you do for a week.